ADVANCED CURRICULUM FROM THE
CENTER FOR GIFTED EDUCATION AT WILLIAM & MARY

MATH Curriculum *for Gifted Students*

GRADE **3**

Lessons, Activities, and Extensions for Gifted and Advanced Learners

Student Workbook
Sections III-IV

CENTER FOR GIFTED EDUCATION

WITH MARGARET JESS MCKOWEN PATTI

William & Mary
School of Education

CENTER FOR GIFTED EDUCATION

P.O. Box 8795
Williamsburg, VA 23187

First published in 2020 by Prufrock Press Inc.

Published in 2021 by Routledge
605 Third Avenue, New York, NY 10017
2 Park Square, Milton Park, Abingdon, Oxon OX14 4RN

Routledge is an imprint of the Taylor & Francis Group, an informa business.

ISBN-13: 978-1-64632-022-6

Edited by Lacy Compton

Cover and layout design by Allegra Denbo and Shelby Charette

NEW YORK AND LONDON

TABLE OF CONTENTS

LESSON 3.1 ACTIVITY
Fractional Lengths

Directions: Because of your excellent calculation skills, you have been hired to complete a tricky measurement job dealing with the plumbing system at a new construction site. You will work with a partner to measure various straws that represent the pipes to fit the needs of the construction site. One straw (pipe) represents one whole and every fraction that you and your partner work with today will be related to this whole. Make sure to reference this whole when needed, and also make sure that this pipe never gets cut, or you could have a major leak and lose your job! Complete the steps below, using the number line on page 2.

1. In the kitchen, under the sink, the contractor has asked you to make two $\frac{1}{2}$ pieces of pipe.
   a. How many total sections will you have after you mark the pipe with your dry erase marker? _____
   b. If the whole pipe is 8 inches long, how long will each piece be? _____

   c. Now follow the directions to cut the pipe and label the number line.

2. The contractor has asked for some pipes that are $\frac{1}{4}$ of the whole pipe.
   a. How many total sections will you have after you mark the pipe with your dry erase marker? _____
   b. If the whole pipe is about 8 inches long, how long will each piece be? _____

   c. Now follow the directions to cut the pipe and label the number line.

3. To fix an issue in the bathroom sink, you are asked to cut pipe. You need each piece to represent $\frac{1}{5}$ of the whole pipe.
   a. How many total sections will you have after you mark the pipe with your dry erase marker? _____
   b. If the whole pipe is about 8 inches long, how long will each piece be? _____

   c. Now follow the directions to cut the pipe and label the number line.

4. Another issue with the plumbing has caused you to have to cut even smaller pieces of pipe. You now need pipe that is $\frac{1}{6}$ the size of the original pipe.

    a. How many total sections will you have after you mark the pipe with your dry erase marker? _____

    b. If the whole pipe is about 8 inches long, how long will each piece be? _____

    c. Now follow the directions to cut the pipe and label the number line.

5. To fix an outside pipe, the contractor has asked you to created a piece of pipe that is $\frac{1}{8}$ the size of the whole pipe.

    a. How many total sections will you have after you mark the pipe with your dry erase marker? _____

    b. If the whole pipe is about 8 inches long, how long will each piece be? _____

    c. Now follow the directions to cut the pipe and label the number line.

6. You realize that $\frac{1}{8}$ is not the size pipe you needed to fix the outside plumbing problem, so you cut pipe the size of $\frac{1}{10}$ of the whole.

    a. How many total sections will you have after you mark the pipe with your dry erase marker? _____

    b. If the whole pipe is about 8 inches long, how long will each piece be? _____

    c. Now follow the directions to cut the pipe and label the number line.

7. Finally, the last job requires a piece of pipe that is $\frac{1}{12}$ the length of the whole pipe.

    a. How many total sections will you have after you mark the pipe with your dry erase marker? _____

    b. If the whole pipe is about 8 inches long, how long will each piece be? _____

    c. Now follow the directions to cut the pipe and label the number line.

Extend Your Thinking

1. Suppose the job requires the use of a pipe $2\frac{1}{3}$ inches in length. How could you represent this measurement with the straws?

LESSON 3.1 PRACTICE
Fractions and Number Lines

1. The local movie theater is tracking attendance at the shows.
   a. On Thursday, there were originally 18 people at the movie, but 3 people left. What fraction of people left? _____

   b. What fraction of people remained? _____

2. On Friday night, 18 people entered the theater, but 6 people left early.
   a. What fraction of people left? _____

   b. What fraction of people remained? _____

3. Did the movie have better attendance on Thursday night or Friday night? Explain your answer.

4. The movie theater sold pizzas for people to snack on while watching the movie.
   a. The first pizza displayed was pepperoni and was cut into 8 slices. Four pieces were eaten. Draw a number line below, and label the fraction of pieces that were sold.

   b. The next pizza displayed was sausage and was also cut into eighths, but only 2 slices were eaten. On the same number line, label the fraction of pieces that were sold.
   c. Look at the number line and the two fractions. Which pizza sold more pieces? _____

5. Some people bought whole pizzas to eat during the movie. The chicken pizzas were sold in small sizes and large sizes. Jeremiah purchased the small pizza and ate the whole thing. Kerri purchased the large pizza, and she also ate the whole thing. Kerri says she ate the same amount as Jeremiah, but Jeremiah says he ate less. Who do you agree with? Explain your answer.

Extend Your Thinking

1. Determine what fraction of your classmates has brown eyes. What fraction has blue eyes? Collect the data. Then represent the data on a number line.

LESSON 3.1
Assessment Practice

Directions: Complete the problems below.

1. Shade $\frac{3}{4}$ of the rectangle.

2. Shade $\frac{1}{3}$ of the whole circle.

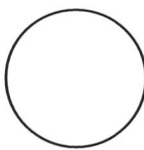

3. Which number is not represented by the picture?

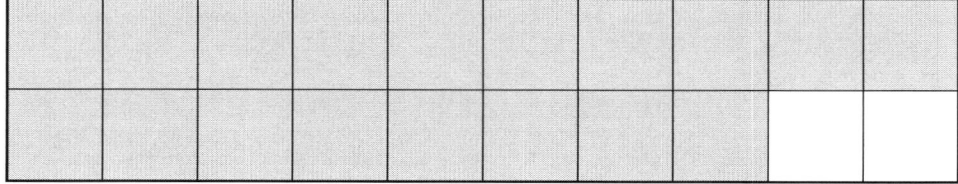

    a. 1.8

    b. $1\frac{8}{10}$

    c. $\frac{18}{10}$

    d. 8

4. Place a dot on the number line where $\dfrac{5}{12}$ would be located. Explain why you placed the dot in that spot.

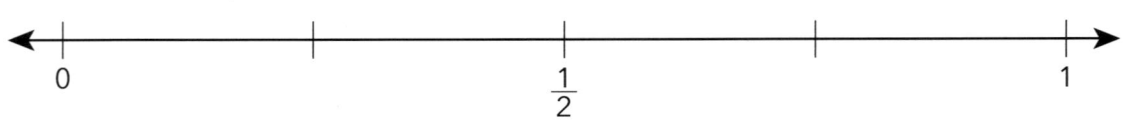

5. Place a dot on the number line where $\dfrac{7}{8}$ would be located. Explain why you placed a dot in that spot.

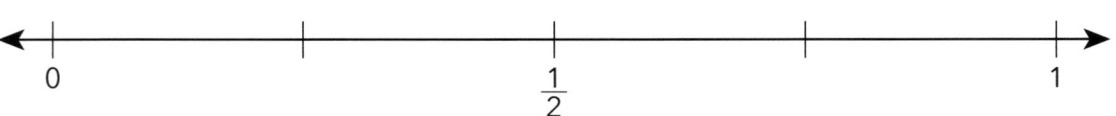

LESSON 3.2 ACTIVITY

Equivalence

Directions: Fractions can be written in numerous ways, depending how each individual thinks about them. Put on your thinking cap, and try to determine several ways to write fractions by working with a partner to find two fraction cards that are equivalent to each of the fractions on the activity sheet.

Once you have found a pair of fractions equivalent to the fraction on the activity sheet, answer the question. That will help you create yet another equivalent fraction. Be sure to discuss with your partner and provide reasoning as to how you knew the fractions were equivalent.

Given Fraction	Equivalent Fraction Card	Equivalent Fraction Card	Create Equivalent Fraction
$\frac{2}{8}$			Divide a box into 24 equal sections. How many sections should be shaded to represent $\frac{2}{8}$?
$\frac{4}{6}$			If 20 boxes were being counted and shaded (numerator), how many total sections would the box have to be broken into (denominator)?

Given Fraction	Equivalent Fraction Card	Create Equivalent Fraction
$\dfrac{3}{10}$ (number line)		If the same length number line was broken into 20 equal sections, where would the tick mark have to be to create an equivalent fraction?
□□□○○○□□○ What fraction of the shapes are circles?		If the denominator of a fraction was 28, what would the numerator have to be to create an equivalent fraction?

Extend Your Thinking

1. Think of a fraction. Now create three fractions that are equivalent to your fraction. Represent each fraction with a picture or on the number line.

2. Think about the mixed number $5\dfrac{3}{4}$. Draw a picture to represent the number. Then locate and label the number on the given number line.

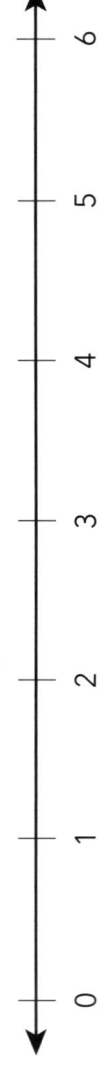

0   1   2   3   4   5   6

Math Curriculum for Gifted Students, Grade 3, Sections III–IV

LESSON 3.2
Fraction Cards

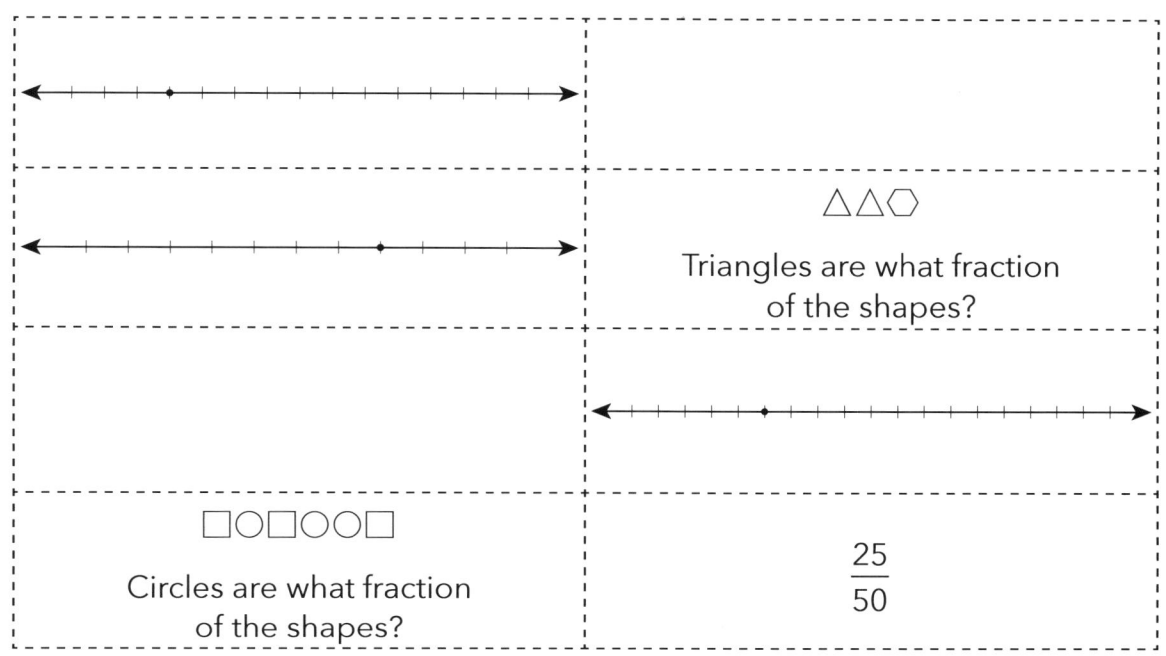

LESSON 3.2 PRACTICE
Equivalent Fractions

Directions: Complete the problems below.

1. Carlie and Ian were celebrating their birthdays. Suppose Carlie cut her cake into 9 pieces as shown below and Ian cut his cake into 5 pieces as shown below. Carlie says to Ian that her birthday cake was bigger, because she was able to cut her cake in to more pieces.
   a. If the pictures below are accurate, how should Ian respond?

   Carlie 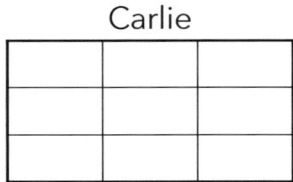   Ian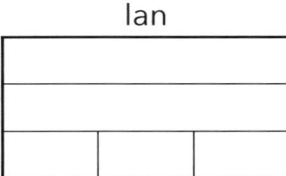

   b. How could you cut the cakes to make them look alike? Draw a model of your answer below.

2. Each pizza at the birthday party was sliced into 9 pieces. Ian ate all of the slices in one pizza.
   a. What fraction of the pizza did he eat? _____

   b. What is another way to represent the amount of pizza Ian consumed?

3. Carlie says that altogether, all of the guests ate 10 whole pizzas. How could you represent that as a fraction if you are actually referring to a number larger than one whole? Explain your fraction.

Extend Your Thinking

1. How does multiplying or dividing the numerator and denominator of a fraction create an equivalent fraction? What is happening? Draw a model to show your thinking.

LESSON 3.2

Assessment Practice

Directions: Complete the problems below.

1. You need to purchase $\frac{3}{4}$ of a yard of fabric from the fabric store. When you arrived, you did not see any fabric pieces labeled $\frac{3}{4}$ of a yard. Which of the following labels is equivalent to $\frac{3}{4}$ yard?

   a. $\frac{1}{2}$

   b. $\frac{6}{8}$

   c. $\frac{3}{9}$

   d. $\frac{1}{4}$

2. Explain how you know that the piece of fabric you chose is equivalent to $\frac{3}{4}$ yard.

3. You bought four pieces of fabric. The employee cuts the fabric into four equal parts from a larger piece of fabric (see the picture below). Write two different fractions that show the part of the bigger piece of fabric that you bought.

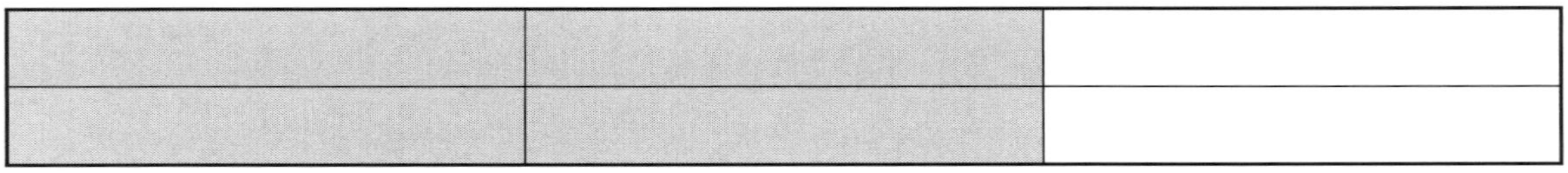

4. From the fabric store, you also needed to purchase $\frac{5}{8}$ yard of yarn. You locate yarn labeled $\frac{5}{8}$ inch. Is this the correct amount of yarn that you should purchase? Why or why not?

5. Draw a picture that shows how 42 divided by 2 is 21.

LESSON 3.3 ACTIVITY
Comparing Fractions

Directions: Who can roll the largest or smallest fraction? Let's find out! You and a partner will work together to compare fractions by playing a dice game.

In the first round, Partner A will roll Die A, and Partner B will decide the probability of rolling a larger fraction using the same die. Partner B will then attempt to roll a larger fraction. Discuss with each other who rolled the larger fraction and by how many eighths. Record your work on the chart and answer the questions as you play. Continue until the two of you each roll three times.

For the second round, Partner B will roll first. This time, after Partner B rolls Die B, Partner A will determine the probability of rolling a smaller fraction. Partner A will then roll the die. Discuss with each other who rolled the smaller fraction and how you know. Record your work on the chart and answer the questions as you play. Continue until the two of you each roll three times.

Extend Your Thinking

1. Make your own dice with fractions of unlike numerators and denominators, and have your classmates compare them by using their knowledge of multiples.

LESSON 3.3
Comparing Fractions Chart

1. Play the first round with Die A and fill in the chart as you play.

Partner A's Roll	Chances of Rolling a Larger Fraction	Partner B's Roll	Who Rolled the Larger Fraction?	How Many Eighths Larger?
1				
2				
3				
4				
5				
6				

2. Order all of your combined rolls from least to greatest. Draw a number line and represent each roll on the number line. Duplicate rolls do not need to be recorded more than once.

3. Play the second round with Die B and fill in the chart as you play.

Partner A's Roll	Chances of Rolling a Smaller Fraction	Partner B's Roll	Who Rolled the Smaller Fraction?	How Do You Know?
1				
2				
3				
4				
5				
6				

LESSON 3.3
Die A

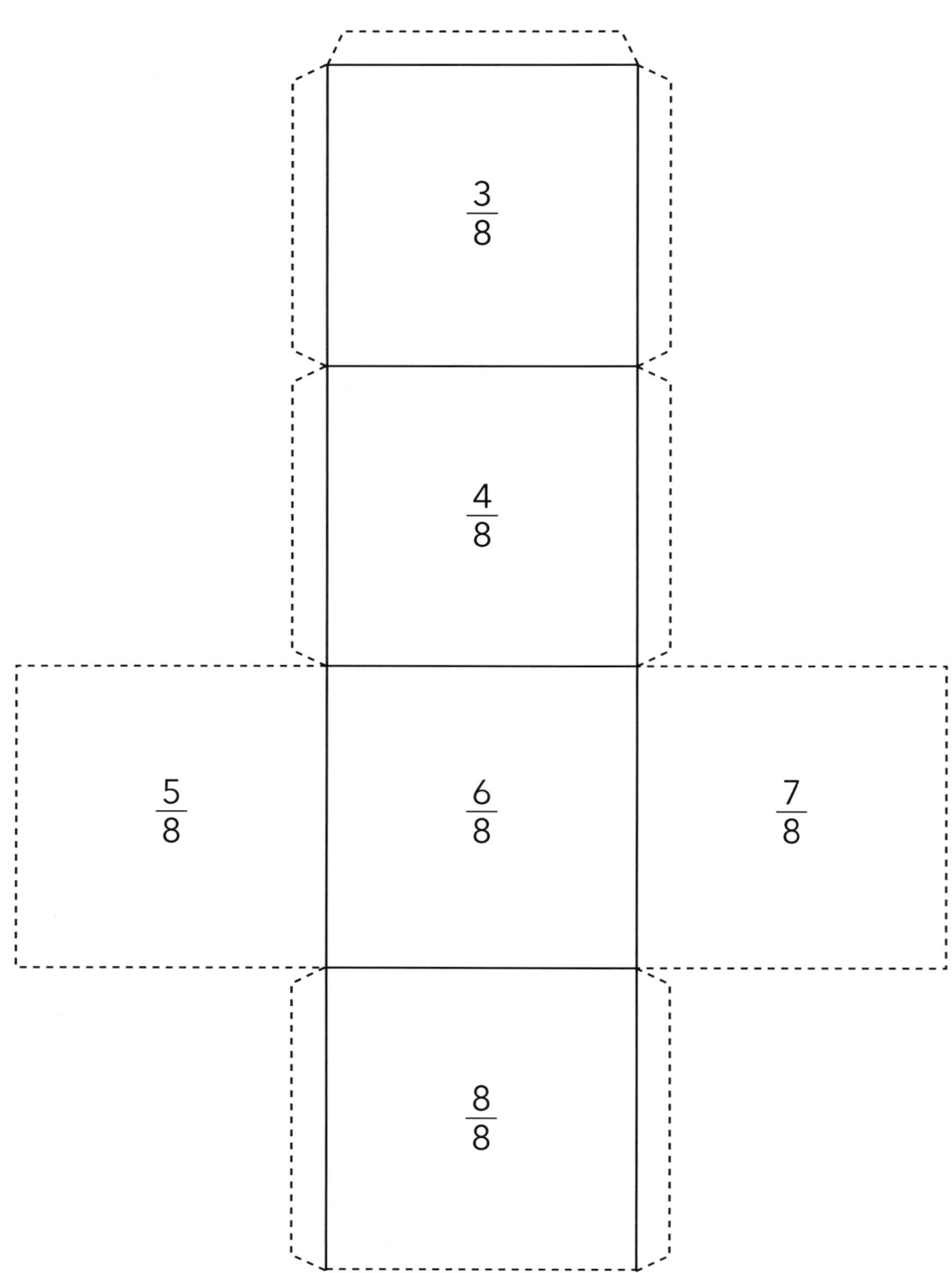

LESSON 3.3
Die B

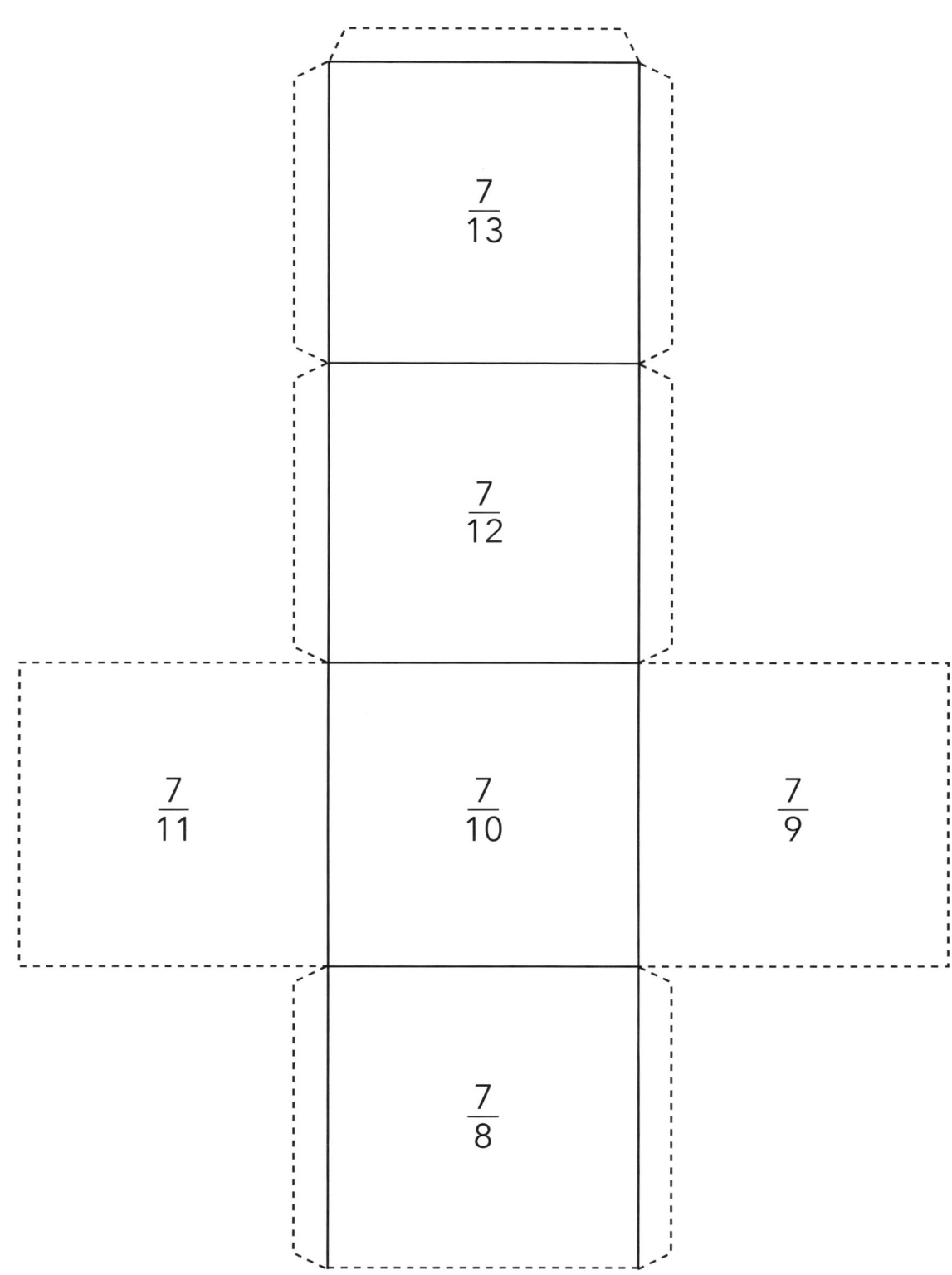

LESSON 3.3 PRACTICE
Comparing Fractions

Directions: Complete the problems below.

1. Marvin was disappointed because he only received $\frac{1}{5}$ of a pie and says that his friend Peter received more pie because his dessert was cut from the same size pie, but he got to eat $\frac{1}{7}$ of the pie. Who ate more pie? Explain using pictures and words.

2. Half of Mrs. Oats's class of 20 students and half of Mrs. Haynes's class of 26 students made A honor roll. The students say that the classes tied because half of each class scored all A's. Are the students correct? Explain.

3. Make a generalization about comparing fractions with common denominators that are referring to the same whole.

4. Make a generalization about comparing fractions with numerators that are alike but do not have the same whole.

LESSON 3.3

Assessment Practice

Directions: Complete the problems below.

1. Which fraction is smaller than $\frac{7}{10}$?

   a. $\frac{9}{10}$

   b. $\frac{6}{10}$

   c. $\frac{8}{10}$

   d. $\frac{11}{10}$

2. The local movie shop is having a sale. Of the movies that Claudia buys, $\frac{4}{5}$ are drama and $\frac{4}{7}$ were comedy. Which is the larger fraction? How do you know?

3. Which sequence of fractions is ordered from least to greatest?

   a. $\frac{2}{5}, \frac{1}{2}, \frac{1}{5}$

   b. $\frac{7}{8}, \frac{7}{9}, \frac{9}{9}$

   c. $\frac{4}{8}, \frac{4}{7}, \frac{4}{9}$

   d. $\frac{2}{9}, \frac{2}{8}, \frac{2}{5}$

4. Which fraction is larger than $\frac{3}{5}$?

    a. $\frac{3}{4}$

    b. $\frac{3}{7}$

    c. $\frac{3}{11}$

    d. $\frac{3}{13}$

LESSON 4.1 ACTIVITY
Classifying Shapes

Directions: Shapes are all around us, and some shapes are even made up of other shapes. You and a partner will work together to classify shapes. Use the bag of shapes and sort them into different categories in each Venn diagram below. Then, sketch the shapes on the cards onto the Venn diagram. Be sure to answer the questions that follow each diagram.

1. Fill in the labeled Venn diagram with the correct shapes.

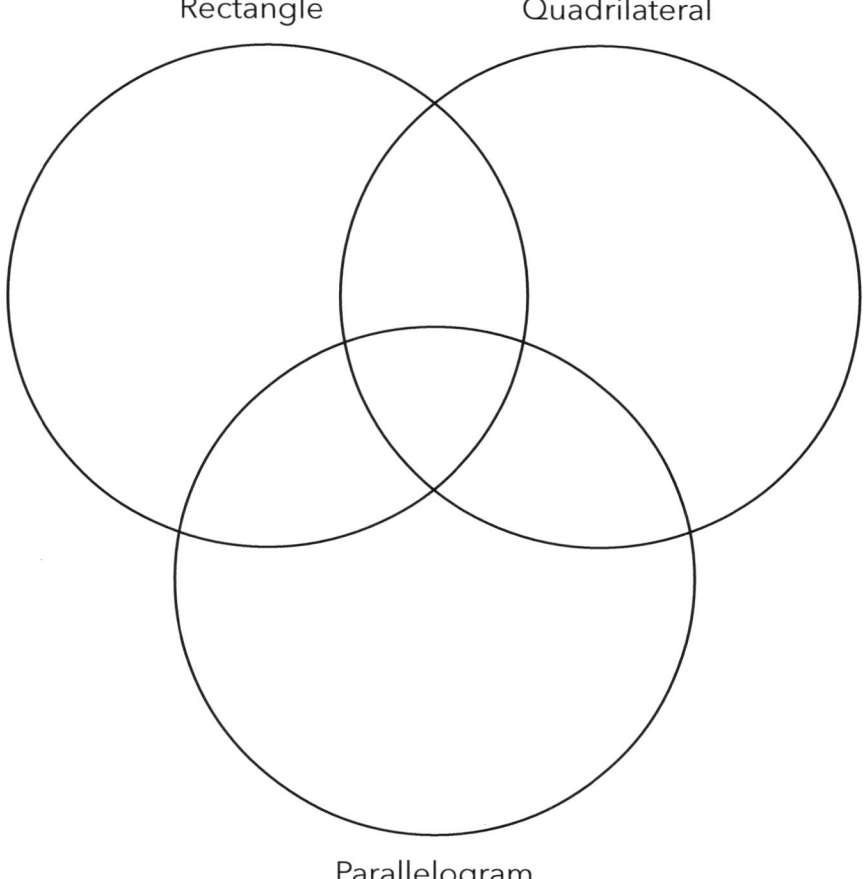

Rectangle   Quadrilateral

Parallelogram

2. What generalization can you make about parallelograms after completing this Venn diagram?

3. Fill in the Venn diagram with the correct shapes.

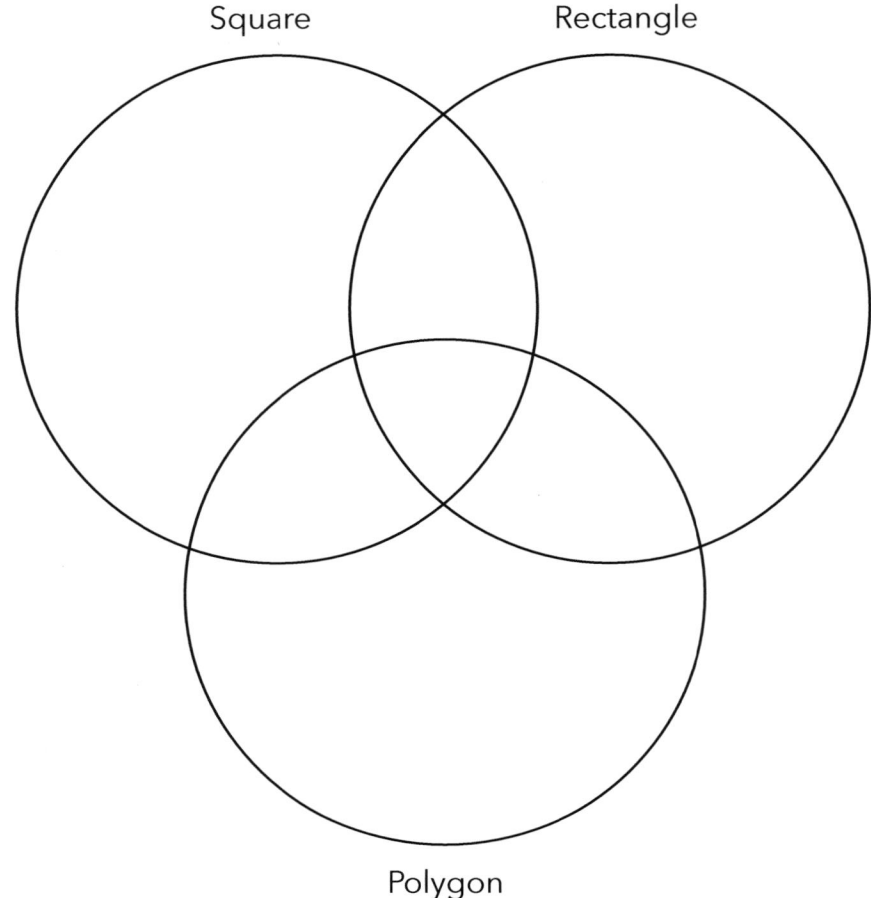

Square   Rectangle

Polygon

4. Of the provided shapes, which is the only one to be classified as a square, rectangle, and a polygon?

5. Fill in the Venn diagram with the correct shapes.

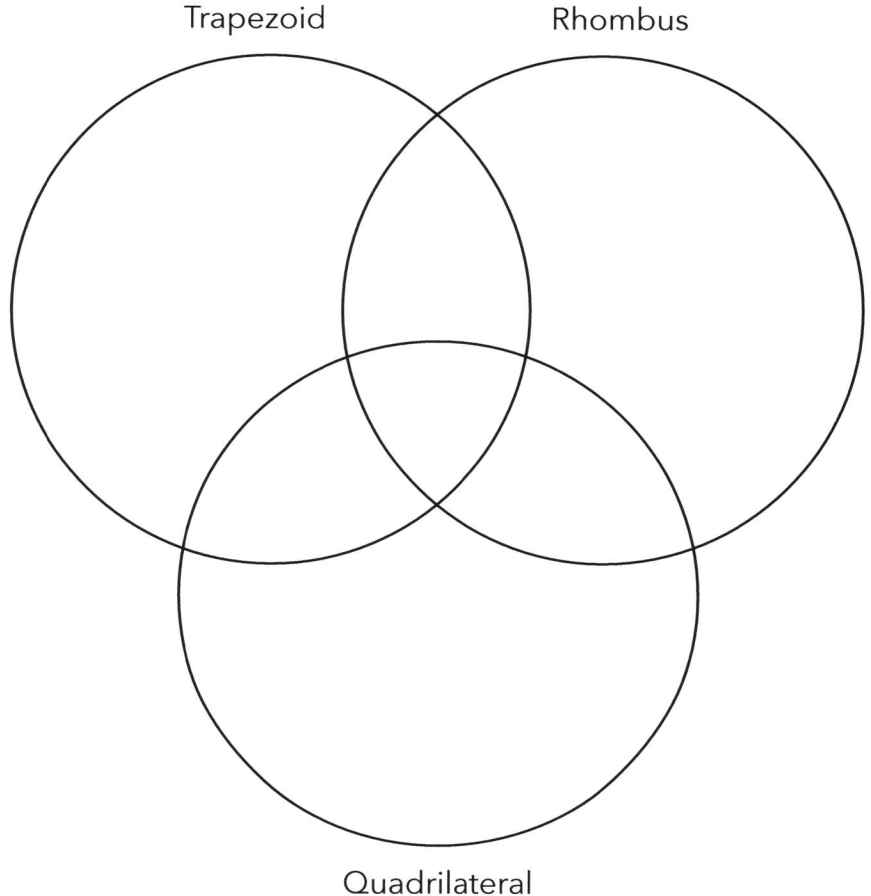

Trapezoid   Rhombus

Quadrilateral

6. Explain why none of the shapes fall in the center of the Venn diagram.

7. Fill in the Venn diagram with the correct shapes.

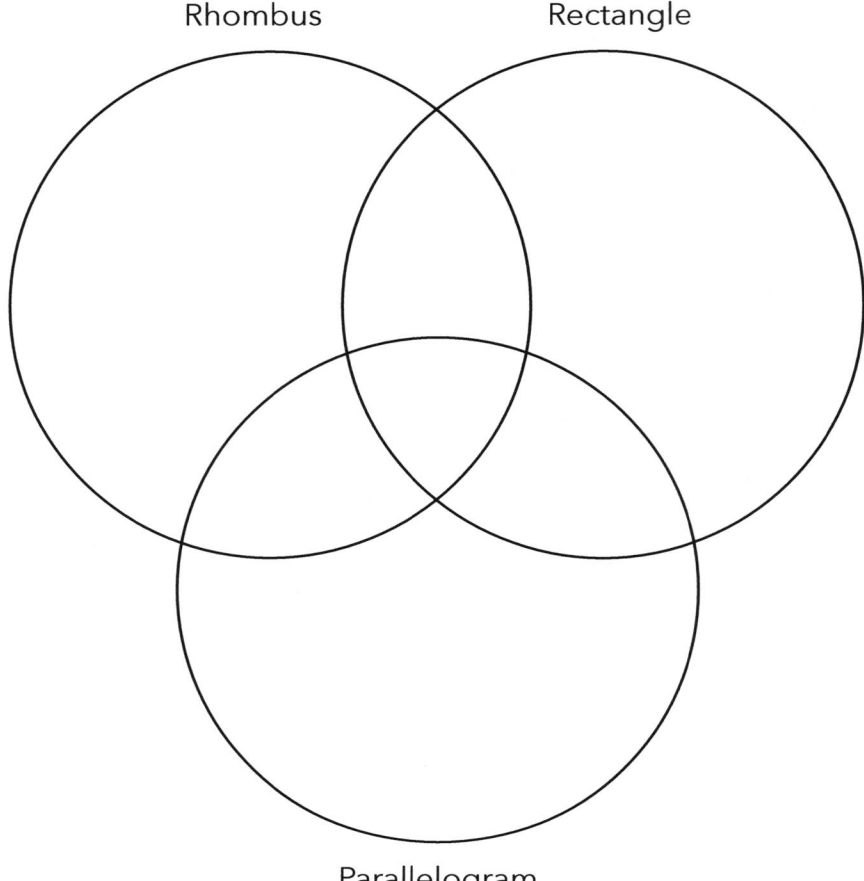

Rhombus   Rectangle

Parallelogram

Extend Your Thinking

1. Create a Venn diagram and place the shapes in the correct places, but do not include the labels for the Venn diagram. Ask your partner to analyze the shapes in each section and determine the labels for each.

LESSON 4.1
Shapes

LESSON 4.1 PRACTICE

Classifying Shapes

Directions: Complete the problems below.

1. This shape can be classified in several ways. Name five categories in which a square fits.

2. What is the broadest category that all of the shapes in this activity fit? All of the shapes we worked with today are _____.

3. Draw a shape that is a quadrilateral but is not a rhombus, a rectangle, or a square.

4. Draw a rectangle and a square. Compare and contrast the two figures.

5. Describe the attributes of a trapezoid.

6. Study the placement of the shapes below. Decide on the label for each section of the Venn diagram. Write in the labels.

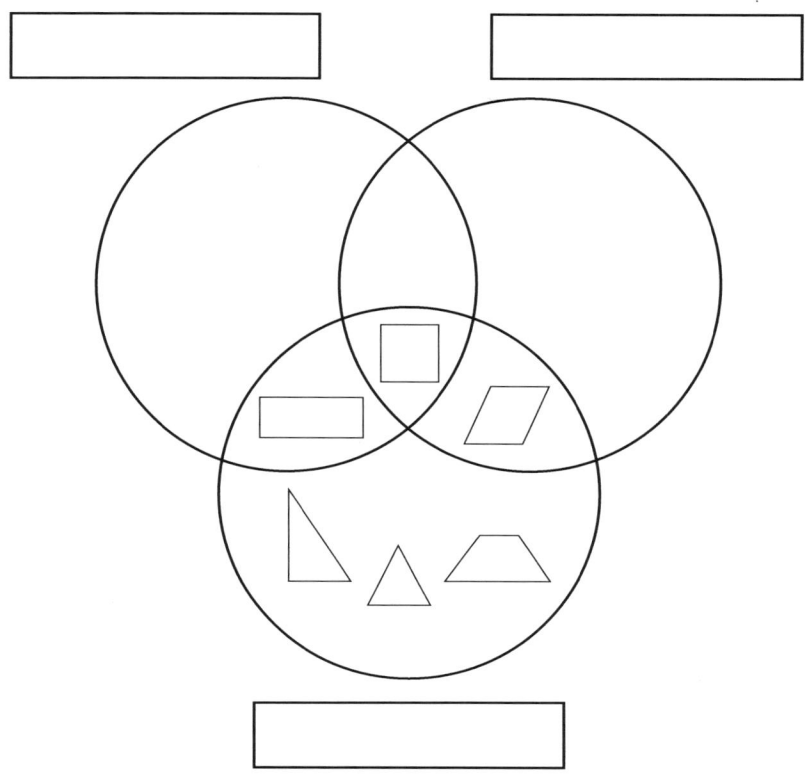

Extend Your Thinking

1. Create a tree model as a visual representation of the categories of shapes.

LESSON 4.1

Assessment Practice

Directions: Complete the problems below.

1. Rectangles and parallelograms have similar and unlike characteristics.
   a. Compare and contrast a rectangle and a parallelogram.

   b. Draw two figures that demonstrate the difference between the two terms.

2. Which shape is a rhombus?

   a.

   b.

   c.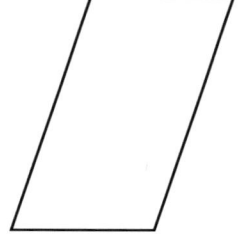

   d.

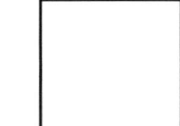

Math Curriculum for Gifted Students, Grade 3, Sections III–IV

Section IV: Geometry

3. Classify the shape. Be as specific as you can.

4. Why isn't the shape in Number 3 a parallelogram? After explaining, extend lines on the picture to show why it isn't a parallelogram.

LESSON 4.2 ACTIVITY
Partitioning

Directions: Different shapes have different areas, but their areas are still related to one another. You and your partner will search for the card with the triangle that is labeled "First Card." Whoever finds it first will be Partner A. Partner A will read the card aloud, and Partner B will look through the other cards to find the card being described. Discuss with each other to decide if you agree upon the card choice. Continue play with Partner B reading the chosen card and Partner A finding the referenced card. Remember to place the cards back because they can be used more than once.

Once the game is complete, use the tangram shapes to help you fill out the chart below. List the first shape, state how its area is related to the second shape, and then draw a picture to model the relationship.

Shape 1	Area Relationship	Shape 2	Picture

Extend Your Thinking

1. Use tangram shapes to create a story about partitioning the area of shapes.

LESSON 4.2
Shape Cards

First Card

Triangle

My area is $\frac{1}{6}$ of whose area?

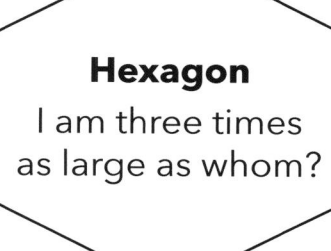

Hexagon

I am three times as large as whom?

Rhombus

Whose area is $\frac{1}{2}$ of my area?

Square

My area is $\frac{1}{3}$ of whose area?

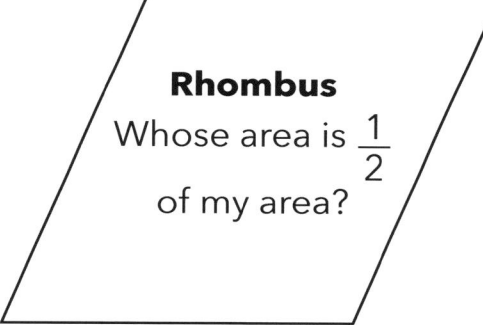

Trapezoid

I am $\frac{1}{2}$ of whose area?

LESSON 4.2 PRACTICE
Partitioning Shapes

Directions: Complete the problems below.

1. Consider the fractional pieces and draw the whole.

a.
$$\frac{1}{6}$$

c.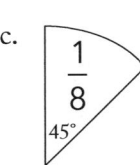
$$\frac{1}{8}$$
45°

b.
$$\frac{1}{4}$$

d.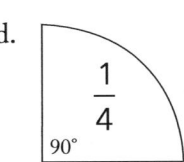
$$\frac{1}{4}$$
90°

2. Name an item in your classroom that has an area about $\frac{1}{6}$ the area of your desktop.
_____

3. Partition the following shapes to show the size of the fractional piece listed.

Shape	Fractional Piece	Drawing of Fractional Piece
	$\frac{3}{8}$	
	$\frac{1}{2}$	

Shape	Fractional Piece	Drawing of Fractional Piece
	$\frac{1}{4}$	Draw two different ways.
	$\frac{1}{4}$	Draw three different ways.

Extend Your Thinking

1. Trace the shapes in the table on Number 3 onto a clean sheet of paper. Partition each shape into a different number of sections.
   a. How many different ways are possible? _____
   b. Can each shape be broken into the same number of equal sections? _____

LESSON 4.2

Assessment Practice

1. Draw a square and partition it into six parts. Shade three parts.

    a. What fraction of the shape is shaded? _____

    b. What fraction of the shape is not shaded? _____

    c. Why do the shaded and nonshaded portions on the shape have the same fraction?

2. For the shape to be partitioned into equal sections, estimate how many more lines need to be added.

    a. 8

    b. 2

    c. 7

    d. 3

3. Samantha says that there is no need to add more lines to the above shape. Explain to Samantha why more lines must be added in order to determine what fractional amount has been shaded.

4. What fractional part has been shaded?

    a. $\dfrac{2}{3}$

    b. $\dfrac{5}{6}$

    c. $\dfrac{4}{4}$

    d. $\dfrac{4}{7}$

For Product Safety Concerns and Information, please contact our EU representative: GPSR@taylorandfrancis.com Taylor & Francis Verlag GmbH, Kaufingerstraße 24, 80331 München, Germany.

PGIL2024USA

First published in 2020 by Prufrock Press Inc.

Published in 2021 by Routledge
605 Third Avenue, New York, NY 10017
2 Park Square, Milton Park, Abingdon, Oxon OX14 4RN

Routledge is an imprint of the Taylor & Francis Group, an informa business.

ISBN-13: 978-1-64632-022-6

Edited by Lacy Compton

Cover and layout design by Allegra Denbo and Shelby Charette

Routledge
Taylor & Francis Group
NEW YORK AND LONDON

LESSON 3.1 ACTIVITY
Fractional Lengths

Directions: Because of your excellent calculation skills, you have been hired to complete a tricky measurement job dealing with the plumbing system at a new construction site. You will work with a partner to measure various straws that represent the pipes to fit the needs of the construction site. One straw (pipe) represents one whole and every fraction that you and your partner work with today will be related to this whole. Make sure to reference this whole when needed, and also make sure that this pipe never gets cut, or you could have a major leak and lose your job! Complete the steps below, using the number line on page 2.

1. In the kitchen, under the sink, the contractor has asked you to make two $\frac{1}{2}$ pieces of pipe.
   a. How many total sections will you have after you mark the pipe with your dry erase marker? _____
   b. If the whole pipe is 8 inches long, how long will each piece be? _____

   c. Now follow the directions to cut the pipe and label the number line.

2. The contractor has asked for some pipes that are $\frac{1}{4}$ of the whole pipe.
   a. How many total sections will you have after you mark the pipe with your dry erase marker? _____
   b. If the whole pipe is about 8 inches long, how long will each piece be? _____

   c. Now follow the directions to cut the pipe and label the number line.

3. To fix an issue in the bathroom sink, you are asked to cut pipe. You need each piece to represent $\frac{1}{5}$ of the whole pipe.
   a. How many total sections will you have after you mark the pipe with your dry erase marker? _____
   b. If the whole pipe is about 8 inches long, how long will each piece be? _____

   c. Now follow the directions to cut the pipe and label the number line.

Math Curriculum for Gifted Students, Grade 3, Sections III–IV

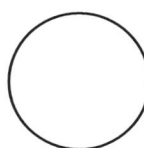

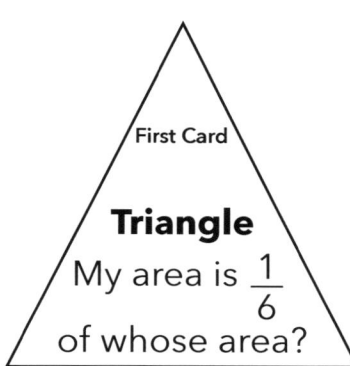

LESSON 4.2 PRACTICE
Partitioning Shapes

Directions: Complete the problems below.

1. Consider the fractional pieces and draw the whole.

   a. $\dfrac{1}{6}$

   b. $\dfrac{1}{4}$

   c. $\dfrac{1}{8}$ 45°

   d. $\dfrac{1}{4}$ 90°

2. Name an item in your classroom that has an area about $\dfrac{1}{6}$ the area of your desktop.
   _____

3. Partition the following shapes to show the size of the fractional piece listed.

Shape	Fractional Piece	Drawing of Fractional Piece
	$\dfrac{3}{8}$	
	$\dfrac{1}{2}$	

Math Curriculum for Gifted Students, Grade 3, Sections III–IV

Section III: Number and Operations–Fractions

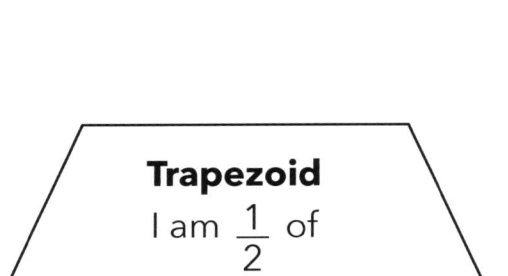

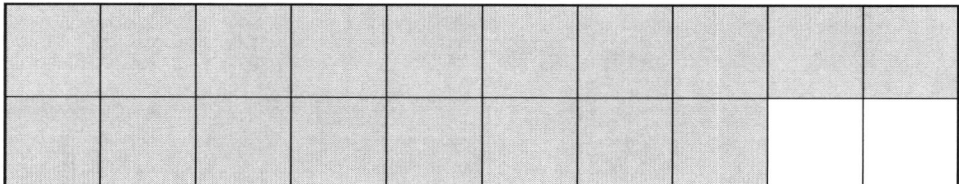

LESSON 3.3
Die A

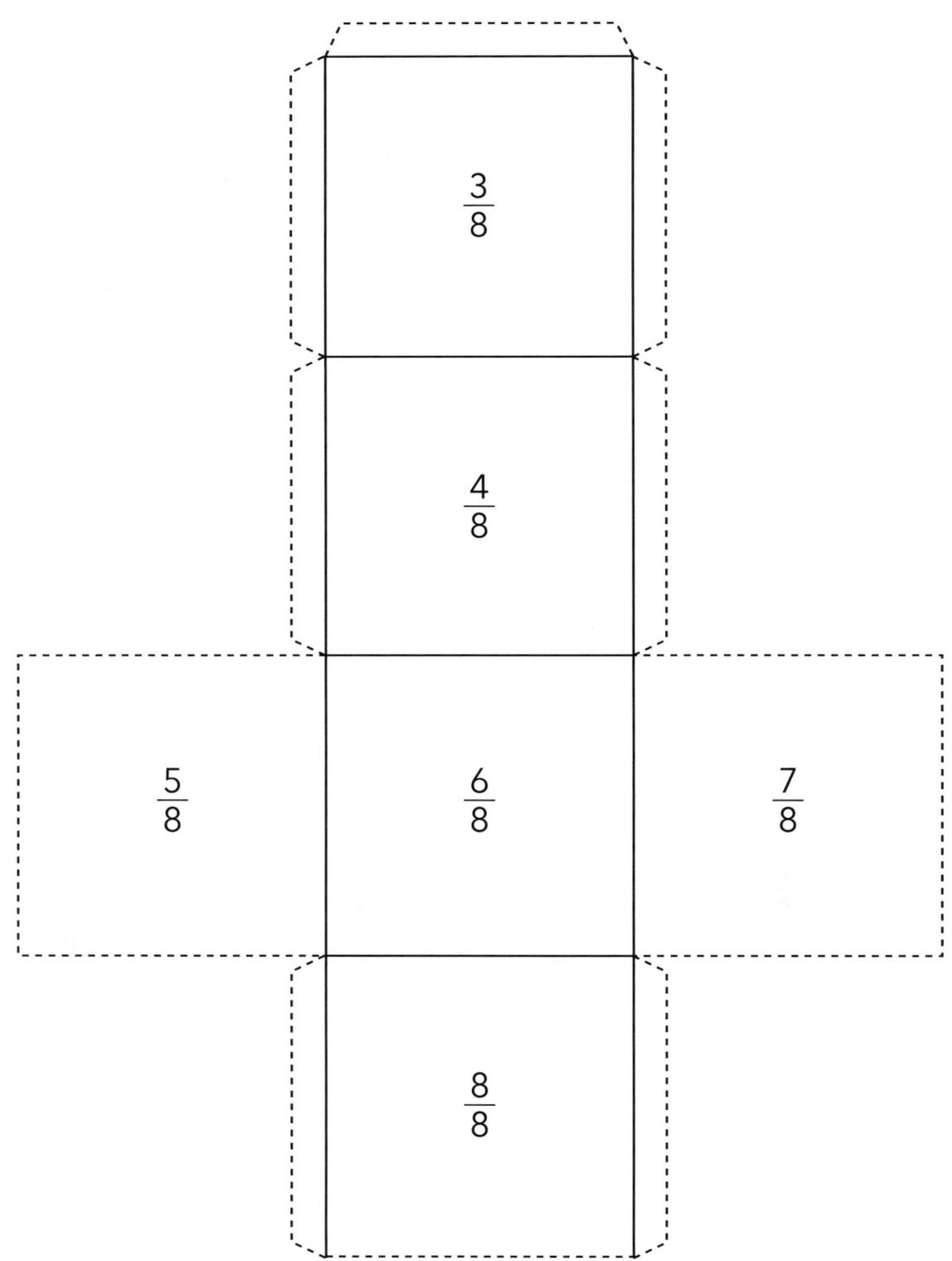

Math Curriculum for Gifted Students, Grade 3, Sections III–IV

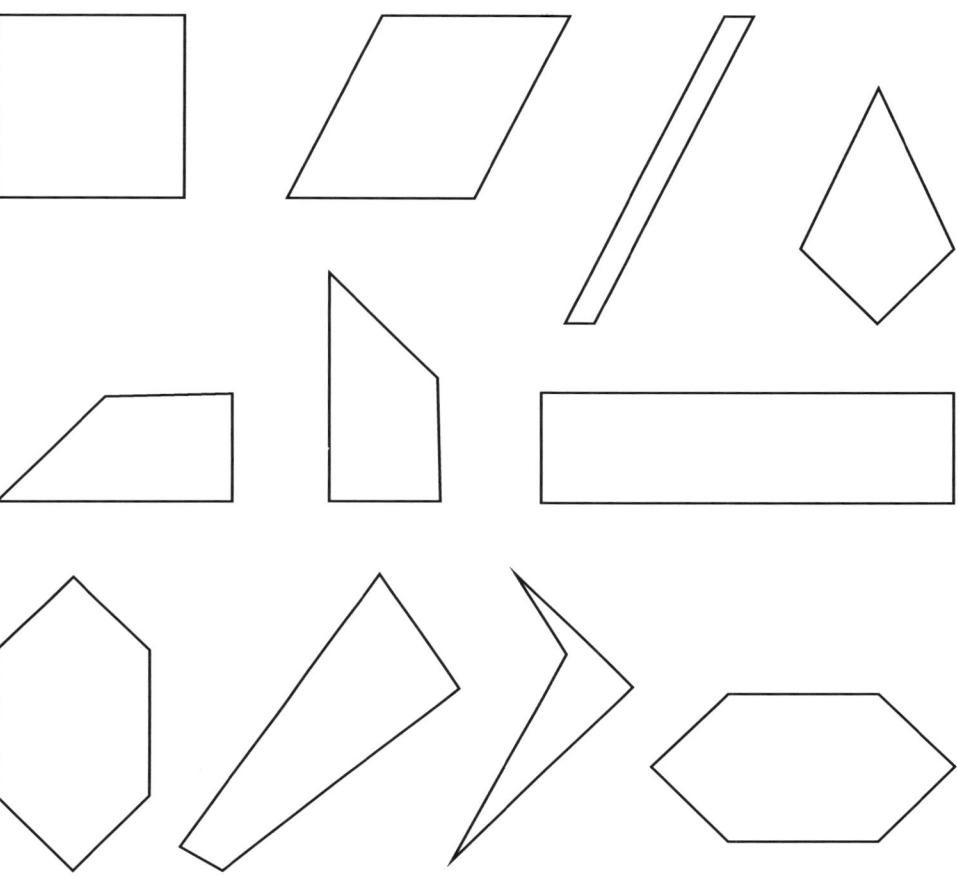

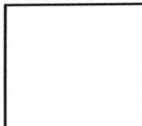

   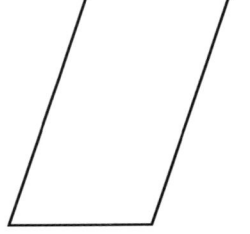

